The
American
West as
Living
Space

WALLACE
STEGNER

. . . .

The American West as Living Space

The University of
Michigan Press
Ann Arbor

Copyright © by The University of Michigan 1987

All rights reserved

Published in the United States of America by

The University of Michigan Press

Manufactured in the United States of America

1993 1992 1991 1990   8 7 6 5

Library of Congress Cataloging-in-Publication Data

Stegner, Wallace Earle, 1909–
   The American West as living space.

   "Derives with only minor changes from a series of
three William W. Cook Lectures delivered at the Law
School of the University of Michigan in Ann Arbor
on October 28, 29, and 30, 1986"—Pref.
   Bibliography: p.
   1. West (U.S.)—Description and travel. 2. Natural
resources—West (U.S.) 3. Environmental protection—
West (U.S.) I. Title.
F591.S823   1987   917.8   87-19114
ISBN 0-472-09375-4 (alk. paper)
ISBN 0-472-06375-8 (pbk. : alk. paper)

Photographs courtesy of The Wilderness Society

PREFACE

. . . .

THIS book derives with only minor changes
from a series of three William W. Cook Lectures delivered at the Law School of the University of
Michigan in Ann Arbor on October 28, 29, and 30, 1986.

The subject to be discussed was the West. A sensible
way to discuss it would have been to select some manage-
able aspect of it and focus on that—to discuss the West
geographically as high plains, Rocky Mountains, southwestern plateaus, Great Basin deserts, the California lit-
toral, and the Pacific Northwest; or historically as the
romantic frontier of the fur trade, the wagon trains, the
mining rushes, and the sod house homesteads; or philo-
sophically as the dream of six-shooter freedom and
orange-grove bliss; or economically as the New West of
sunshine cities, the energy boom, and the grandiose,
costly, federally funded effort that has converted some
western valleys into gardens and most western rivers into
plumbing systems; or sociologically as an emergent sub-
species of the half-defined American culture.

That would have been orderly. It would also have been
the way the blind men approached the elephant. I decided
that I would rather risk superficiality and try to leave an
impression of the region in all its manifestations, to try a

holistic portrait, a look at the gestalt, the whole shebang, than settle for a clear impression of some treelike, spearlike, or ropelike part.

It used to be a boast among travelers in the West that they had "seen the elephant." I do not flatter myself that in three one-hour lectures I have been able to do more than sketch an outline. But I hope that it is at least an outline of the whole thing, the living region, with its country and its people, its splendors and its limitations, its facts and its fantasies, its opportunities and its problems, its romantic past and booming present and dubious future all suggested.

I have painted with a broad brush because there was no space to do more; and in the end I have concluded that the space limitation was salutary: it made me concentrate upon the essentials and kept me from getting tangled up in detail. And I have been personal because the West is not only a region but a state of mind, and both the region and the state of mind are my native habitat.

CONTENTS

. . . .

The
American
West as
Living
Space

. . . .

ONE

Living Dry

THE West is a region of extraordinary variety within its abiding unity, and of an iron immutability beneath its surface of change. The most splendid part of the American habitat, it is also the most fragile. It has been misinterpreted and mistreated because, coming to it from earlier frontiers where conditions were not unlike those of northern Europe, we found it different, daunting, exhilarating, dangerous, and unpredictable, and we entered it carrying habits that were often inappropriate, and expectations that were surely excessive. The dreams we brought to it were recognizable American dreams—a new chance, a little gray home in the West, adventure, danger, bonanza, total freedom from constraint and law and obligation, the Big Rock Candy Mountain, the New Jerusalem. Those dreams had often paid off in earlier parts of America, and they paid off for some in the West. For the majority, no. The West has had a way of warping well-carpentered habits, and raising the grain on exposed dreams.

The fact is, it has been as notable for mirages as for the realization of dreams. Illusion and mirage have been built into it since Coronado came seeking the Seven Cities of Cíbola in 1540. Coronado's failure was an early, spectacular trial run for other and humbler failures. Witness the young men from all over the world who fill graveyards in California's Mother Lode country. There is one I remember: *Nato a Parma 1830, morto a Morfi 1850,* an inscription as significant for its revelation of the youth of many argonauts as for its misspelling of Murphy's, the camp where this boy died. Witness too the homesteaders who retreated eastward from the dry plains with signs on their covered wagons: "In God we trusted, in Kansas we busted." Yet we have not even yet fully lost our faith in Cíbola, or the nugget as big as a turnip, or even Kansas.

Anyone pretending to be a guide through wild and fabulous territory should know the territory. I wish I knew it better than I do. I am not Jed Smith. But Jed Smith was not available for this assignment, and I was. I accepted it eagerly, at least as much for what I myself might learn as for what I might be able to tell others. I can't come to even tentative conclusions about the West without coming to some conclusions about myself.

I have lived in the West, many parts of it, for the best part of seventy-seven years. I have found stories and novels in it, have studied its history and written some of it, have tried to know its landscapes and understand its people, have loved and lamented it, and sometimes rejected its most "western" opinions and prejudices, and pretty consistently despised its most powerful politicians and the general trend of their politics. I have been a lover but not much of a booster. Nevertheless, for better or worse, the West is in my computer, the biggest part of my software.

If there is such a thing as being conditioned by climate and geography, and I think there is, it is the West that has conditioned me. It has the forms and lights and colors that I respond to in nature and in art. If there is a western speech, I speak it; if there is a western character or personality, I am some variant of it; if there is a western culture in the small-*c*, anthropological sense, I have not escaped it. It has to have shaped me. I may even have contributed to it in minor ways, for culture is a pyramid to which each of us brings a stone.

Therefore I ask your indulgence if I sometimes speak in terms of my personal experience, feelings, and values, and put the anecdotal and normative ahead of the statistical, and emphasize personal judgments and trial syntheses rather than the analysis that necessarily preceded them. In doing so, I shall be trying to define myself as well as my native region.

Perhaps we will all know better what I think when we see what I say.

There are other ways of defining the West, but since Major John Wesley Powell's 1878 *Report on the Lands of the Arid Region* it has usually been said that it starts about the 98th meridian of west longitude and ends at the Pacific Ocean. Neither boundary has the Euclidean perfection of a fixed imaginary line, for on the west the Pacific plate is restless, constantly shoving Los Angeles northward where it is not wanted, and on the east the boundary between Middle West and West fluctuates a degree or two east or west depending on wet and dry cycles.

Actually it is not the arbitrary 98th meridian that marks the West's beginning, but a perceptible line of real import that roughly coincides with it, reaching southward about a third of the way across the Dakotas, Nebraska, and Kansas, and then swerving more southwestward across Oklahoma and Texas. This is the isohyetal line of twenty inches, beyond which the mean annual rainfall is less than the twenty inches normally necessary for unirrigated crops.

A very little deficiency, even a slight distortion of the season in which the rain falls, makes all the difference. My family homesteaded on the Montana-Saskatchewan border in 1915, and burned out by 1920, after laying the foundation for a little dust bowl by plowing up a lot of buffalo grass. If the rains had been kind, my father would have proved up on that land and become a naturalized Canadian. I estimate that I missed becoming Canadian by no more than an inch or two of rain; but that same deficiency confirmed me as a citizen of the West.

The West is defined, that is, by inadequate rainfall, which means a general deficiency of water. We have water only between the time of its falling as rain or snow and the time when it flows or percolates back into the sea or the deep subsurface reservoirs of the earth. We can't create water, or increase the supply. We can only hold back and redistribute what there is. If rainfall is inadequate, then streams will be inadequate, lakes will be few and sometimes saline, underground water will be slow to renew itself when it has been pumped down, the air will be very dry, and surface evaporation from lakes and reservoirs will be extreme. In desert parts of the West it is as much as ten feet a year.

The only exception to western aridity, apart from the mountains that provide the absolutely indispensable snowsheds, is the northwest corner, on the Pacific side of the Cascades. It is a narrow exception: everything east of the mountains, which means two-thirds to three-quarters of Washington and Oregon, is in the rain shadow.

California, which might seem to be an exception, is not.

Utah, 1940

Though from San Francisco northward the coast gets plenty of rain, that rain, like the lesser rains elsewhere in the state, falls not in the growing season but in winter. From April to November it just about can't rain. In spite of the mild coastal climate and an economy greater than that of all but a handful of nations, California fits Walter Webb's definition of the West as "a semi-desert with a desert heart." It took only the two-year drought of 1976–77, when my part of California got eight inches of rain each year instead of the normal eighteen, to bring the whole state to a panting pause. A five-year drought of the same severity would half depopulate the state.

So—the West that we are talking about comprises a dry core of eight public lands states—Arizona, Colorado, Idaho, Montana, Nevada, New Mexico, Utah, and Wyoming—plus two marginal areas. The first of these is the western part of the Dakotas, Nebraska, Kansas, Oklahoma, and Texas, authentically dry but with only minimal public lands. The second is the West Coast—Washington, Oregon, and California—with extensive arid lands but with well-watered coastal strips and with many rivers. Those marginal areas I do not intend to exclude, but they do complicate statistics. If I cite figures, they will often be for the states of the dry core.

A RIDITY, and aridity alone, makes the various Wests one. The distinctive western plants and animals, the hard clarity (before power plants and metropolitan traffic altered it) of the western air, the look and location of western towns, the empty spaces that separate them, the way farms and ranches are either densely concentrated where water is plentiful or widely scattered where it is scarce, the pervasive presence of the federal government as landowner and land manager, the even more noticeable federal presence as dam builder and water broker, the snarling states'-rights and antifederal feelings whose burden Bernard DeVoto once characterized in a sentence—"Get out and give us more money"—those are all consequences, and by no means all the consequences, of aridity.

Aridity first brought settlement to a halt at the edge of the dry country and then forced changes in the patterns of settlement. The best guide to those changes is Walter Webb's seminal study, *The Great Plains* (1931). As Webb pointed out, it took a lot of industrial invention to conquer the plains: the Colt revolver, a horseman's weapon, to subdue the horse Indians; barbed wire to control cattle; windmills to fill stock tanks and irrigate little gardens and hayfields; railroads to open otherwise unlivable spaces and bring first buffalo hides and buffalo bones and then cattle and wheat to market; gang machinery to plow, plant, and harvest big fields.

As it altered farming methods, weapons, and tools, so the dry country bent water law and the structure of land ownership. Eastern water law, adopted with little change from English common law, was essentially riparian. The owner of a stream bank, say a miller, could divert water from the stream to run his mill, but must flow it back when it had done his work for him. But in the arid lands only a little of the water diverted from a stream for any purpose—gold mining, irrigation, municipal or domestic consumption—ever finds its way back. Much is evaporated and lost. Following the practice of Gold Rush miners who diverted streams for their rockers and monitors, all of the dry core states subscribe to the so-called Colorado Doctrine of prior appropriation. First come, first served. The three coastal states go by some version of the California Doctrine, which is modified riparian.

Because water is precious, and because either prior appropriation or riparian rights, ruthlessly exercised, could give monopolistic power to an upstream or riparian landowner, both doctrines, over time, have been hedged with safeguards. Not even yet have we fully adapted our water law to western conditions, as was demonstrated when the California Supreme Court in 1983 invoked the "Public Trust" principle to prevent a single user, in this case the Los Angeles Water and Power Authority, from ever again draining a total water supply, in this case the streams that once watered the Owens Valley and fed Mono Lake. The court did not question the validity of Los Angeles' water rights. It only said that those rights could not be exercised at the expense of the public's legitimate interest.

The conditions that modified water use and water law also changed the character of western agriculture. The open range cattle outfits working north from Texas after the Civil War had only a brief time of uninhibited freedom before they ran into nesters and barbed wire. But the homesteaders who began disputing the plains with cattlemen quickly found that 160 acres of dry land were a hand without even openers. Many threw in. Those who stuck either found land on streams or installed windmills; and they moved toward larger acreages and the stock farm, just as the cattlemen eventually moved toward the same adjustment from the other end.

Powell had advocated just such an adjustment in 1878. In country where it took 20, 30, even 50 acres to feed one steer, quarter-section homesteads were of no use to a stock raiser. But 160 acres of intensively worked irrigated land were more than one family needed or could handle. Furthermore, the rectangular cadastral surveys that had been in use ever since the Northwest Ordinance of 1787 paid no attention to water. Out on the plains, a single quarter or half section might contain all the water within miles, and all the adjacent range was dominated by whoever owned that water.

Powell therefore recommended a new kind of survey defining irrigable homesteads of 80 acres and grazing homesteads of 2,560, four full sections. Every plot should have access to water, and every water right should be tied to land title. Obviously, that program would leave a lot of dry land unsettled.

Unfortunately for homesteaders and the West, Powell's report was buried under dead leaves by Congress. Ten years later, his attempt to close the Public Domain until he could get it surveyed and its irrigable lands identified was defeated by Senator "Big Bill" Stewart of Nevada, the first of a long line of incomparably bad Nevada senators.

In 1889 and 1890 the constitutional conventions of Montana, Idaho, and Wyoming territories would not listen when Powell urged them to lay out their political boundaries along drainage divides, so that watershed and timber lands, foothill grazing lands, and valley irrigated lands could be managed intelligently without conflict. The only place where a drainage divide does mark a political boundary is a stretch of the Continental Divide between Idaho and Montana. And to cap this history of old habits stubbornly clung to and hopes out of proportion to possibilities, when Powell addressed the boosters of the Irrigation Congress in Los Angeles in 1893 and warned them that they were laying up a heritage of litigation and failure because there was only enough water to irrigate a fifth of the western lands, the boosters didn't listen either. They booed him.

Powell understood the consequences of aridity, as the boosters did not, and still do not. Westerners who would like to return to the old days of free grab, people of the kind described as having made America great by their initiative and energy in committing mass trespass on the minerals, grass, timber, and water of the Public Domain, complain that no western state is master in its own house. Half its land is not its own. 85 percent of Nevada is not Nevada, but the United States; two-thirds of Utah and Idaho likewise; nearly half of California, Arizona, and Wyoming—48 percent of the eleven public lands states.

There are periodic movements, the latest of which was the so-called Sagebrush Rebellion of the 1970s, to get these lands "returned" to the states, which could then dispose of them at bargain-basement rates to favored stockmen, corporations, and entrepreneurs.

The fact is, the states never owned those lands, and gave up all claim to them when they became states. They were always federal lands, acquired by purchase, negotiation, or conquest before any western state existed. The original thirteen colonies created the first Public Domain when they relinquished to the federal government their several claims to what was then the West. The states between the Alleghanies and the Mississippi River were made out of it. All but one of the rest of the contiguous forty-eight were made out of the Louisiana and Florida purchases, the land acquired by the Oregon settlement with Great Britain, and the territory taken from Mexico in the Mexican War or obtained by purchase afterward. All of the western states except Texas, which entered the Union as an independent republic and never had any Public Domain, were created out of federal territory by formal acts of Congress, which then did everything it could to dispose of the public lands within them.

As far as a little beyond the Missouri, the system of disposal worked. Beyond the 98th meridian it did not, except in the spotty way that led Webb to call the West an "oasis civilization." Over time, large areas of forest land and the most spectacular scenery were reserved in the public interest, but much land was not considered worth reserving, and could not be settled or given away. The land laws—the Preemption Act, Homestead Act, Desert Land Act, Carey Act, Timber and Stone Act—produced more failure and fraud than family farms. Not even the Newlands Act of 1902, though it has transformed the West, has put into private hands more than a modest amount of the Public Domain.

Despite all efforts, the West remained substantially federal. In 1930 Herbert Hoover and his Secretary of the Interior Ray Lyman Wilbur tried to give a lot of abused dry land to the states, and the states just laughed. Then in 1934, in the worst Dust Bowl year, the Taylor Grazing Act acknowledged the federal government's reluctant decision to retain—and rescue and manage—the overgrazed, eroded Public Domain. In the 1940s the stockmen's associations tried to steal it, as well as the Forest Service's grazing lands, in a dry rehearsal of the later Sagebrush Rebellion. Bernard DeVoto, in the Easy Chair pages of *Harper's Magazine,* almost single-handedly frustrated that grab.

But by the 1940s more than stockmen were interested, and more than grass was at stake. During and after World War II the West had revealed treasures of oil, coal, uranium, molybdenum, phosphates, and much else. States whose extractive industries were prevented from unhampered exploitation of these resources again cried foul, demanding "back" the lands they had never owned, cared for, or really wanted, and complaining that the acreages of the Bureau of Land Management, national forests, national parks and monuments, wildlife refuges, military reservations, and dam sites, as well as the arid poor farms where we had filed away our Indian responsibilities, were off the state tax rolls and outside of state control and "locked up" from developers who, given a free hand, would make the West rich and prosperous.

Never mind that grazing fees and coal and oil lease fees and timber sale prices are so low that they amount to a fat subsidy to those who enjoy them. (The flat fee of $1.35 per Animal-Unit-Month set by the Reagan Administration as a grazing fee is about 20 percent of what leasers of private land must pay in most districts. Many timber sales are below cost). Never mind that half of the money that does come in from fees and leases is given to the states in lieu of taxes. Never mind that the feds spend the other half, and more, rescuing and rehabilitating the lands whose proper management would bankrupt the states in which they lie. Never mind the federal aid highways, and the federally financed dams, and the write-offs against flood control, and the irrigation water delivered at a few dollars an acre foot. Take for granted federal assistance, but damn federal control. Your presence as absentee landlord offends us, Uncle. Get out, and give us more money.

There are other objections, too, some of them more legitimate than those of stockmen, lumbermen, and miners greedy for even more than they are getting. There are real difficulties of management when the landscape is checkerboarded with private, state, and federal owner-ship. And federal space is tempting whenever the nation needs a bombing range, an atomic test site, missile silo locations, or places to dump nuclear waste or store nerve gas. Generally, the West protests that because of its public lands it gets all the garbage; but sometimes, so odd we are as a species, one or another western state will fight for the garbage, and lobby to become the home of a nuclear dump.

More by oversight than intention, the federal government allowed the states to assert ownership of the water within their boundaries, and that is actually an ownership far more valuable but more complicated than that of land. The feds own the watersheds, the stream and lake beds, the dam sites. Federal bureaus, with the enthusiastic concurrence of western chambers of commerce, have since 1902 done most of the costly impoundment and distribution of water. And federal law, in a pinch, can and does occasionally veto what states and irrigation districts do with that water. It is a good guess that it will have to do so more and more—that unless the states arrive at some relatively uniform set of rules, order will have to be imposed on western water by the federal government.

It will not be easy, and the federal government has created a nasty dilemma for itself by giving its blessing both to the legal fiction of state water ownership and to certain Native American rights to water. As early as 1908, in the so-called Winters Doctrine decision, the Supreme Court confirmed the Indians' rights in water originating within or flowing through their reservations. Though those rights have never been quantified or put to use (a beginning has been made by the tribes on the Fort Peck Reservation), Indian tribes all across the West have legitimate but unspecified claims to water already granted by the states to white individuals and corporations. Even without the Indian claims, many western streams—

for a prime example the Colorado—are already over-subscribed. The Colorado River Compact allocated 17.5 million acre feet annually among the upper basin, lower basin, and Mexico. The actual annual flow since 1930 has averaged about 12 million acre feet.

Aridity arranged all that complicated natural and human mess, too. In the view of some, it also helped to create a large, spacious, independent, sunburned, self-reliant western character, and a large, open, democratic western society. Of that, despite a wistful desire to believe, I am less than confident.

NINETEENTH-CENTURY America, overwhelmingly agricultural, assumed that settlement meant agricultural settlement. That assumption underlies—some say it undermines—Frederick Jackson Turner's famous hypothesis, expressed in his 1893 lecture "The Significance of the Frontier in American History," that both democratic institutions and the American character have been largely shaped by the experience of successive frontiers, with their repeated dream of betterment, their repeated acceptance of primitive hardships, their repeated hope and strenuousness and buoyancy, and their repeated fulfillment as smiling and productive commonwealths of agrarian democrats.

Turner was so intent upon the mass movement toward free land that he paid too little attention to a growing movement toward the industrial cities, and his version of the American character is therefore open to some qualifications. He also paid too little attention to the drastic

17

changes enforced by aridity beyond the 98th meridian, the changes that Powell before him and Webb after him concentrated upon.

The fact is, agriculture at first made little headway in the West, and when it was finally imposed upon the susceptible—and some unsusceptible—spots, it was established pretty much by brute force, and not entirely by agrarian democrats.

Until the Civil War and after, most of the West was not a goal but a barrier. Webb properly remarks that if it had turned out to be country adapted to the slave economy, the South would have fought for it, and its history would have been greatly different. He also points out that if the country beyond the Missouri had been wooded and well watered, there would have been no Oregon Trail.

Emigrants bound up the Platte Valley on their way to Oregon, California, or Utah, the first targets of the westward migration, almost universally noted in their journals that a little beyond Grand Island their nostrils dried out and their lips cracked, their wagon wheels began to shrink and wobble, and their estimates of distance began to be ludicrously off the mark. They observed that green had ceased to be the prevailing color of the earth, and had given way to tans, grays, rusty reds, and toned white; that salt often crusted the bottoms of dry lakes (they used some of it, the sodium or potassium bicarbonate that they called saleratus, to leaven their bread); that the grass no longer made a turf, but grew in isolated clumps with bare earth between them; that there was now no timber except on the islands in the Platte; that unfamiliar animals had appeared: horned toads and prairie dogs that seemed to require no water at all, and buffalo, antelope, jackrabbits, and coyotes that could travel long distances for it.

They were at the border of strangeness. Only a few miles into the West, they felt the difference; and as Webb says, the degree of strangeness can be measured by the fact that almost all the new animals they saw they misnamed. The prairie dog is not a dog, the horned toad is not a toad, the jackrabbit is not a rabbit, the buffalo is not a buffalo, and the pronghorn antelope is more goat than antelope. But they could not mistake the aridity. They just didn't know how much their habits would have to change if they wanted to live beyond the 98th meridian. None of them in that generation would have denied aridity to its face, as William Gilpin, the first territorial governor of Colorado, would do a generation later, asserting that it was a simple matter on the plains to *dig* for wood and water, that irrigation was as easy as fencing, which it supplanted, and that the dry plains and the Rockies could handily support a population of two hundred million. That kind of fantasy, like Cyrus Thomas's theory that cultivation increases rainfall, that "rain follows the plow," would have to wait for the boosters.

Two lessons all western travelers had to learn: mobility and sparseness. Mobility was the condition of life beyond the Missouri. Once they acquired the horse, the Plains Indians were as migrant as the buffalo they lived by. The mountain men working the beaver streams were no more fixed than the clouds. And when the change in hat fashions killed the fur trade, and mountain men turned to guiding wagon trains, the whole intention of those trains was to get an early start, as soon as the grass greened up, and then get *through* the West as fast as possible. The Mormons were an exception, a special breed headed for sanctuary in the heart of the desert, a people with a uniquely cohesive social order and a theocratic discipline that made them better able to survive.

But even the Mormons were villages on the march, as mobile as the rest until, like Moses from Pisgah's top, they looked upon Zion. Once there, they quickly made themselves into the West's stablest society. But notice: Now, a hundred and forty years after their hegira, they have managed to put only about 3 percent of Utah's land under cultivation; and because they took seriously the Lord's command to be fruitful and multiply, Zion has been overpopulated, and exporting manpower, for at least half a century. One of the bitterest conflicts in modern Utah is that between the environmentalists who want to see much of that superlative wilderness preserved roadless and wild, and the stubborn Mormon determination to make it support more Saints than it possibly can.

LIEUTENANT Zebulon Pike, sent out in 1806 to explore the country between the Missouri and Santa Fe, had called the high plains the Great American Desert. In 1819 the expedition of Major Stephen Long corroborated that finding, and for two generations nobody seriously questioned it. The plains were unfit for settlement by a civilized, meaning an agricultural, people, and the farther west you went, the worse things got. So for the emigrants who in 1840 began to take wheels westward up the Platte Valley, the interior West was not a place but a way, a trail to the Promised Land, an adventurous, dangerous rite of passage. In the beginning there were only two *places* on its two-thousand-mile length: Fort Laramie on the North Platte and Fort Hall on the Snake. And those were less settlements than way stations, refreshment and recruitment stops different only in style from motel-and-gas-and-lunchroom turnouts on a modern interstate.

Insofar as the West was a civilization at all between the time of Lewis and Clark's explorations and about 1870, it was largely a civilization in motion, driven by dreams. The people who composed and represented it were part of a true Folk-Wandering, credulous, hopeful, hardy, largely uninformed. The dreams are not dead even today, and the habit of mobility has only been reinforced by time. If, as Wendell Berry says, most Americans are not placed but displaced persons, then western Americans are the most displaced persons of all.

Ever since Daniel Boone took his first excursion over Cumberland Gap, Americans have been wanderers. When Charles Dickens, in the Mississippi Valley, met a full-sized dwelling house coming down the road at a round trot, he was looking at the American people head-on. With a continent to take over and Manifest Destiny to goad us, we could not have avoided being footloose. The initial act of emigration from Europe, an act of extreme, deliberate disaffiliation, was the beginning of a national habit.

It should not be denied, either, that being footloose has always exhilarated us. It is associated in our minds with escape from history and oppression and law and irksome obligations, with absolute freedom, and the road has always led west. Our folk heroes and our archetypal literary figures accurately reflect that side of us. Leatherstocking, Huckleberry Finn, the narrator of *Moby Dick,* all are orphans and wanderers: any of them could say, "Call me Ishmael." The Lone Ranger has no dwelling place except the saddle. And when teenagers run away these days, in the belief that they are running toward freedom, they more often than not run west. Listen to the Haight-Ashbury dialogues in Joan Didion's *Slouching toward Bethlehem.* Examine the American character as it is self-described in *Habits of the Heart,* by Robert Bellah and others.

But the rootlessness that expresses energy and a thirst for the new and an aspiration toward freedom and personal fulfillment has just as often been a curse. Migrants deprive themselves of the physical and spiritual bonds that develop within a place and a society. Our migratoriness has hindered us from becoming a people of communities and traditions, especially in the West. It has robbed us of the gods who make places holy. It has cut off individuals and families and communities from memory and the continuum of time. It has left at least some of us with a kind of spiritual pellagra, a deficiency disease, a hungering for the ties of a rich and stable social order. Not only is the American home a launching pad, as Margaret Mead said; the American community, especially in the West, is an overnight camp. American individualism, much celebrated and cherished, has developed without its essential corrective, which is belonging. Freedom, when found, can turn out to be airless and unsustaining. Especially in the West, what we have instead of place is space. Place is more than half memory, shared memory. Rarely do Westerners stay long enough at one stop to share much of anything.

The principal invention of western American culture is the motel, the principal exhibit of that culture the automotive roadside. The principal western industry is tourism, which is not only mobile but seasonal. Whatever it might want to be, the West is still primarily a series of brief visitations or a trail to somewhere else; and western literature, from *Roughing It* to *On the Road*, from *The Log of a Cowboy* to *Lonesome Dove*, from *The Big Rock Candy Mountain* to *The Big Sky*, has been largely a literature not of place but of motion.

Trying to capture America in a sentence, Gertrude Stein said, "Conceive a space that is filled with moving." If she had been reared in Boston she might not have seen it so plainly; but she was reared in Oakland. She knew that few Westerners die where they were born, that most live out their lives as a series of uprootings.

ADAPTATION is the covenant that all successful organisms sign with the dry country. Mary Austin, writing of the Owens Valley in the rain shadow of the Sierra, remarked that "the manner of the country makes the usage of life there, and the land will not be lived in except in its own fashion. The Shoshones live like their trees, with great spaces between. . . ."

She might have added, though I don't remember that she did, how often Shoshonean place names contain the syllable -pah: Tonopah, Ivanpah, Pahrump, Paria. In the Shoshonean language, -pah means water, or water hole. The Pah-Utes are the Water Utes, taking their name from their rarest and most precious resource. They live mainly in Utah and Nevada, the two driest states in the Union, and in those regions water is safety, home, life, *place*. All around those precious watered places, forbidding and un-livable, is only space, what one must travel through be-tween places of safety.

Thoroughly adapted, the Pah-Utes (Paiutes) were mi-gratory between fixed points marked by seasonal food supplies and by water. White Americans, once they began to edge into the dry country from east, west, and south, likewise established their settlements on dependable water, and those towns have a special shared quality, a family resemblance.

For the moment, forget the Pacific Coast, furiously bent on becoming Conurbia from Portland to San Diego. For-get the metropolitan sprawl of Denver, Phoenix, Tucson, Albuquerque, Dallas–Fort Worth, and Salt Lake City, growing to the limits of their water and beyond, like bac-terial cultures overflowing the edges of their agar dishes and beginning to sicken on their own wastes. If we want characteristic western towns we must look for them, para-doxically, beyond the West's prevailing urbanism, out in the boondocks where the interstates do not reach, main-line planes do not fly, and branch plants do not locate. The towns that are most western have had to strike a balance between mobility and stability, and the law of sparseness has kept them from growing too big. They are the places where the stickers stuck, and perhaps were stuck; the places where adaptation has gone furthest.

Whether they are winter wheat towns on the subhumid edge, whose elevators and bulbous silver water towers announce them miles away, or county towns in ranch country, or intensely green towns in irrigated desert valleys, they have a sort of forlorn, proud rightness. They look at once lost and self-sufficient, scruffy and indispensable. A road leads in out of wide emptiness, threads a fringe of service stations, taverns, and a motel or two, widens to a couple of blocks of commercial buildings, some still false-fronted, with glimpses of side streets and green lawns, narrows to another strip of automotive roadside, and disappears into more wide emptiness.

The loneliness and vulnerability of those towns always moves me, for I have lived in them. I know how the world of a child in one of them is bounded by weedy prairie, or the spine of the nearest dry range, or by flats where plugged tin cans lie rusting and the wind has pasted paper and plastic against the sagebrush. I know how precious is the safety of a few known streets and vacant lots and familiar houses. I know how the road in both directions both threatens and beckons. I know that most of the children in such a town will sooner or later take that road, and that only a few will take it back.

In mining country, vulnerability has already gone most or all of the way to death. In those ghosts or near ghosts where the placers have been gutted or the lodes played out, the shafts and drifts and tailings piles, the saloons and stores and hotels and houses, have been left to the lizards and a few survivors who have rejected the command of mobility. The deader the town, the more oppressive the emptiness that surrounds and will soon reclaim it. Unless, of course, the federal government has installed an atomic testing site, or new migrant entrepreneurs have come in to exploit the winter skiing or bring summer-festival culture. Then the ghost will have turned, at least temporarily, into a Searchlight, an Aspen, a Telluride, a Park City—a new way station for new kinds of migrants.

WE return to mobility and the space that enforces it. Consider the observations of William Least Heat Moon, touring the blue highways of America. "The true West," he says (and notice that he too finds the true West somewhere outside the cities where 75 percent of Westerners live),

> differs from the East in one great, pervasive,
> influential, and awesome way: space. The vast
> openness changes the roads, towns, houses, farms,
> crops, machinery, politics, economics, and, natu-
> rally, ways of thinking. . . . Space west of the line
> is perceptible and often palpable, especially when it
> appears empty, and it's that apparent emptiness
> which makes matter look alone, exiled, and uncon-

nected. . . . But as the space diminishes man and his constructions in a material fashion it also— paradoxically—makes them more noticeable. *Things show up out here.* The terrible distances eat up speed. Even dawn takes nearly an hour just to cross Texas. (*Blue Highways,* p. 136)

Distance, space, affects people as surely as it has bred keen eyesight into pronghorn antelope. And what makes that western space and distance? The same condition that enforces mobility on all adapted creatures, and tolerates only small or temporary concentrations of human or other life.

Aridity.

And what do you do about aridity, if you are a nation inured to plenty and impatient of restrictions and led westward by pillars of fire and cloud? You may deny it for a while. Then you must either adapt to it or try to engineer it out of existence.

. . . .

TWO

Striking
the Rock

THE summer of 1948 my family and I spent on Struthers Burt's ranch in Jackson Hole. I was just beginning the biography of John Wesley Powell, and beginning to understand some things about the West that I had not understood before. But during that busy and instructive interval my wife and I were also acting as western editors and scouts for a publishing house, and now and then someone came by with a manuscript or the idea for a book. The most memorable of these was a famous architect contemplating his autobiography. One night he showed us slides of some of his houses, including a million-dollar palace in the California desert of which he was very proud. He said it demonstrated that with imagination, technical know-how, modern materials, and enough money, an architect could build anywhere without constraints, imposing his designed vision on any site, in any climate.

In that waterless pale desert spotted with shad scale and creosote bush and backed by barren, lion-colored mountains, another sort of architect, say Frank Lloyd Wright, might have designed something contextual, something low, broad-eaved, thick-walled, something that would mitigate the hot light, something half-underground so that people could retire like the lizards and rattlesnakes from the intolerable daytime temperatures, something made of native stone or adobe or tamped earth in the colors and shapes of the country, something no more visually intrusive than an outcrop.

Not our architect. He had built of cinderblock, in the form of Bauhaus cubes, the only right angles in that desert. He had painted them a dazzling white. Instead of softening the lines between building and site, he had accentuated them, surrounding his sugary cubes with acres of lawn and a tropical oasis of oleanders, hibiscus, and palms—not the native *Washingtonia* palms either, which are a little scraggly, but sugar and royal palms, with a classier, more Santa Barbara look. Water for this estancia, enough water to have sustained a whole tribe of desert Indians, he had brought by private pipeline from the mountains literally miles away.

The patio around the pool—who would live in the desert without a pool?—would have fried the feet of swimmers, three hundred days out of the year, and so he had designed canopies that could be extended and retracted by push button, and under the patio's concrete he had laid pipes through which cool water circulated by day. By night, after the desert chill came on, the circulating water was heated. He had created an artificial climate, inside and out.

Studying that luxurious, ingenious, beautiful, sterile incongruity, I told its creator, sincerely, that I thought he could build a comfortable house in hell. That pleased him; he thought so too. What I didn't tell him, what he would not have understood, was that we thought his desert house immoral. It exceeded limits, it offended our sense not of the possible but of the desirable. There was no economic or social reason for anyone's living on a barren flat, however beautiful, where every form of life sought shelter during the unbearable daylight hours. The only reasons for building there were to let mad dogs and rich men go out in the midday sun, and to let them own and dominate a view they admired. The house didn't fit the country, it challenged it. It asserted America's never-say-never spirit and America's ingenious know-how. It seemed to us an act of arrogance on the part of both owner and architect.

I felt like asking him, What if a super-rich Eskimo wanted a luxury house on Point Hope? Would you build it for him? Would you dam the Kobuk and bring megawatts of power across hundreds of miles of tundra, and set up batteries of blower-heaters to melt the snow and thaw the permafrost, and would you erect an international style house with picture windows through which the Eskimo family could look out across the lawn and strawberry bed and watch polar bears on the pack ice?

He might have taken on such a job, and he was good enough to make it work, too—until the power line blew down or shorted out. Then the Eskimos he had encouraged to forget igloo building and seal oil lamps would freeze into ice sculptures, monuments to human pride. But of course that is all fantasy. Eskimos, a highly adapted and adaptable people, would have more sense than to challenge their arctic habitat that way. Even if they had unlimited money. Which they don't.

That desert house seemed to me, and still seems to me, a paradigm—hardly a paradigm, more a caricature—of what we have been doing to the West in my lifetime. Instead of adapting, as we began to do, we have tried to make country and climate over to fit our existing habits and desires. Instead of listening to the silence, we have

Arizona, 1940

shouted into the void. We have tried to make the arid West into what it was never meant to be and cannot remain, the Garden of the World and the home of multiple millions.

THAT does not mean either that the West should never have been settled or that water should never be managed. The West—the habitable parts of it—is a splendid habitat for a limited population living within the country's rules of sparseness and mobility. If the unrestrained engineering of western water was original sin, as I believe, it was essentially a sin of scale. Anyone who wants to live in the West has to manage water to some degree.

Ranchers learned early to turn creeks onto their hay land. Homesteaders not on a creek learned to dam a runoff coulee to create a "rezavoy" as we did in Saskatchewan in 1915. Kansas and Oklahoma farmers set windmills to pumping up the underground water. Towns brought their water, by ditch or siphon, from streams up on the watershed. Irrigation, developed first by the Southwestern Indians and the New Mexico Spanish, and reinvented by the Mormons—it was a necessity that came with the territory—was expanded in the 1870s and 1880s by such cooperative communities as Greeley, Colorado, and by small-to-medium corporate ventures such as the one I wrote about in *Angle of Repose*—the project on the Boise River that after its failure was taken over by the Bureau of Reclamation and called the Arrowrock Dam.

Early water engineers and irrigators bit off what they

and the local community could chew. They harnessed the streams that they could manage. Some dreamers did take on larger rivers, as Arthur Foote took on the Boise, and went broke at it. By and large, by 1890, individual, corporate, and cooperative irrigators had gone about as far as they could go with water engineering; their modest works were for local use and under local control. It might have been better if the West had stopped there. Instead, all through the 1890s the unsatisfied boosters called for federal aid to let the West realize its destiny, and in 1902 they got the Newlands Act. This *permitted* the feds to undertake water projects—remember that water was state owned, or at least state regulated—and created the Bureau of Reclamation.

Reclamation projects were to be paid for by fees charged irrigation districts, the period for paying off the interest-free indebtedness being first set at ten years. Later that was upped to twenty, later still to forty. Eventually much of the burden of repayment was shifted from the sale of water to the sale of hydropower, and a lot of the burden eliminated entirely by the practice of river-basin accounting, with write-offs for flood control, job creation, and other public goods. Once it was lured in, the federal government—which meant taxpayers throughout the country, including taxpayers in states that resented western reclamation because they saw themselves asked to pay for something that would compete unfairly with their own farmers—absorbed or wrote off more and more of the costs, accepting the fact that reclamation was a continuing subsidy to western agriculture. Even today, when municipal and industrial demands for water have greatly increased, 80 to 90 percent of the water used in the West is used, often wastefully, on fields, to produce crops generally in surplus elsewhere. After all the billions spent by the Bureau of Reclamation, the total area irrigated by its projects is about the size of Ohio, and the water impounded and distributed by the bureau is about 15 percent of all the water utilized in the West. What has been won is only a beachhead, and a beachhead that is bound to shrink.

ONE of the things Westerners should ponder, but generally do not, is their relation to and attitude toward the federal presence. The bureaus administering all the empty space that gives Westerners much of their outdoor pleasure and many of their special privileges and a lot of their pride and self-image are frequently resented, resisted, or manipulated by those who benefit economically from them but would like to benefit more, and are generally taken for granted by the general public.

The federal presence should be recognized as what it is: a reaction against our former profligacy and wastefulness, an effort at adaptation and stewardship in the interest of the environment and the future. In contrast to the principal water agency, the Bureau of Reclamation, which was a creation of the boosters and remains their creature, and whose prime purpose is technological conversion of the arid lands, the land-managing bureaus all have as at least part of their purpose the preservation of the West in a relatively natural, healthy, and sustainable condition.

Yellowstone became the first national park in 1872 because a party of Montana tourists around a campfire voted down a proposal to exploit it for profit, and pledged themselves to try to get it protected as a permanent pleasuring-ground for the whole country. The national forests began because the bad example of Michigan scared Congress about the future of the country's forests, and induced it in 1891 to authorize the reservation of public forest lands by presidential proclamation. Benjamin Harrison took large advantage of the opportunity. Later, Grover Cleveland did the same, and so did Theodore Roosevelt. The West, predictably, cried aloud at having that much plunder removed from circulation, and in 1907 western Congressmen put a rider on an agricultural appropriations bill that forbade any more presidential reservations without the prior consent of Congress. Roosevelt could have pocket vetoed it. Instead, he and Chief Forester Gifford Pinchot sat up all night over the maps and surveys of potential reserves, and by morning Roosevelt had signed into existence twenty-one new national forests, sixteen million acres of them. Then he signed the bill that would have stopped him.

It was Theodore Roosevelt, too, who created the first wildlife refuge in 1903, thus beginning a service whose territories, since passage of the Alaska National Interest Lands Conservation Act of 1980, now exceed those of the National Park Service by ten million acres.

As for the biggest land manager of all, the Bureau of Land Management (BLM), it is the inheritor of the old General Land Office whose job was to dispose of the Public Domain to homesteaders, and its lands are the leftovers once (erroneously) thought to be worthless. Worthless or not, they could not be indefinitely neglected and abused. The health of lands around them depended on their health.

They were assumed as a permanent federal responsibility by the Taylor Grazing Act of 1934, but the Grazing Service then created was a helpless and toothless bureau dominated by local councils packed by local stockmen— foxes set by other foxes to watch the henhouse, in a travesty of democratic local control. The Grazing Service was succeeded by the Bureau of Land Management, which was finally given some teeth by the Federal Land Policy and Management Act of 1976. No sooner did it get the teeth that would have let it do its job than the Sagebrush Rebels offered to knock them out. The Rebels didn't have to. Instead, President Reagan gave them James Watt as secretary of the interior, and James Watt gave them Robert Burford as head of the BLM. The rebels simmered down, their battle won for them by administrative appointment, and BLM remains a toothless bureau.

All of the bureaus walk a line somewhere between preservation and exploitation. The enabling act of the National Park Service in 1916 charged it to provide for the *use without impairment* of the parks. It is an impossible assignment, especially now that more than three hundred million people visit the national parks annually, but the Park Service tries.

The National Forest Service, born out of Pinchot's philosophy of "wise use," began with the primary purpose of halting unwise use, and as late as the 1940s so informed a critic as Bernard DeVoto thought it the very best of the federal bureaus. But it changed its spots during the first Eisenhower administration, under the Mormon patriarch Ezra Taft Benson as secretary of agriculture, and began aggressively to harvest board feet. Other legitimate uses— recreation, watershed and wildlife protection, the gene banking of wild plant and animal species, and especially wilderness preservation—it either neglected or resisted whenever they conflicted with logging.

By now, unhappily, environmental groups tend to see the Forest Service not as the protector of an invaluable public resource and the true champion of multiple use, but as one of the enemy, allied with the timber interests. The Forest Service, under attack, has reacted with a hostility bred of its conviction that it is unjustly criticized. As a consequence of that continuing confrontation, nearly every master plan prepared in obedience to the National Forest Management Act of 1976 has been challenged and will be fought, in the courts if necessary, by the Wilderness Society, the Sierra Club, the Natural Resources Defense Council, and other organizations. The usual charge: too many timber sales, too often at a loss in money as well as in other legitimate values, and far too much roading—roading being a preliminary to logging and a way of forestalling wilderness designation by spoiling the wilderness in advance.

What is taking place is that Congress has been responding to public pressures to use the national forests for newly perceived social goods; and the National Forest Service, for many years an almost autonomous bureau with a high

morale and, from a forester's point of view, high principles, is resisting that imposition of control.

Not even the Fish and Wildlife Service, dedicated to
the preservation of wild species and their habitats, escapes
criticism, for under pressure from stockmen it has historically waged war on predators, especially coyotes, and the
1080 poison baits that it used to distribute destroyed not
only coyotes but hawks, eagles, and other wildlife that the
agency was created to protect. One result has been a good
deal of public suspicion. Even the current device of 1080-
poisoned collars for sheep and lambs, designed to affect
only an attacking predator, is banned in thirty states.

The protection provided by these various agencies is of
course imperfect. Every reserve is an island, and its boundaries are leaky. Nevertheless this is the best protection we
have, and not to be disparaged. All Americans, but especially Westerners whose backyard is at stake, need to ask
themselves whose bureaus these should be. Half of the
West is in their hands. Do they exist to provide bargain-
basement grass to favored stockmen whose grazing privileges have become all but hereditary, assumed and
bought and sold along with the title to the home spread?
Are they hired exterminators of wildlife? Is it their func-
tion to negotiate loss-leader coal leases with energy con-
glomerates, and to sell timber below cost to Louisiana
Pacific? Or should they be serving the much larger public
whose outdoor recreations of backpacking, camping, fishing, hunting, river running, mountain climbing, hang
gliding, and, God help us, dirt biking are incompatible
with clear-cut forests and overgrazed, poison-baited, and

strip-mined grasslands? Or is there a still higher duty—to maintain the health and beauty of the lands they manage, protecting from everybody, including such destructive segments of the public as dirt bikers and pothunters, the watersheds and spawning streams, forests and grasslands, geological and scenic splendors, historical and archaeological remains, air and water and serene space, that once led me, in a reckless moment, to call the western public lands part of the geography of hope?

As I have known them, most of the field representatives of all the bureaus, including the BLM, do have a sense of responsibility about the resources they oversee, and a frequent frustration that they are not permitted to oversee them better. But that sense of duty is not visible in some, and at the moment is least visible in the political appointees who make or enunciate policy. Even when policy is intelligently made and well understood, it sometimes cannot be enforced because of local opposition. More than one Forest Service ranger or BLM man who tried to enforce the rules has had to be transferred out of a district to save him from violence.

There are many books on the Public Domain. One of the newest and best is *These American Lands,* by Dyan Zaslowsky and the Wilderness Society, published in 1986 by Henry Holt and Company. I recommend it, not only for its factual accuracy and clarity, but for its isolation of problems and its suggestions of solutions. Here and now, all I can do is repeat that the land bureaus have a strong, often disregarded influence on how life is lived in the West. They provide and protect the visible, available, unfenced space that surrounds almost all western cities and towns—surrounds them as water surrounds fish, and is their living element.

The bureaus need, and some would welcome, the kind of public attention that would force them to behave in the long-range public interest. Though I have been involved in controversies with some of them, the last thing I would want to see is their dissolution and a return to the policy of disposal, for that would be the end of the West as I have known and loved it. Neither state ownership nor private ownership—which state ownership would soon become—could offer anywhere near the disinterested stewardship that these imperfect and embattled federal bureaus do, while at the same time making western space available to millions. They have been the strongest impediment to the careless ruin of what remains of the Public Domain, and they will be necessary as far ahead as I, at least, can see.

The Bureau of Reclamation is something else. From the beginning, its aim has been not the preservation but the remaking—in effect the mining—of the West.

A principal justification for the Newlands Act was that fabled Jeffersonian yeoman, the small freehold farmer, who was supposed to benefit from the Homestead Act, the Desert Land Act, the Timber and Stone Act, and other land-disposal legislation, but rarely did so west of the 98th meridian. The publicized purpose of federal reclamation was the creation of family farms that would eventually feed the world and build prosperous rural commonwealths in deserts formerly fit for nothing but horned toads and rattlesnakes. To insure that these small farmers would not be done out of their rights by large landowners and water users, Congress wrote into the act a clause limiting the use of water under Reclamation Bureau dams to the amount that would serve a family farm of 160 acres.

Behind the pragmatic, manifest-destinarian purpose of pushing western settlement was another motive: the hard determination to dominate nature that historian Lynn White, in the essay "Historical Roots of Our Ecologic Crisis," identified as part of our Judeo-Christian heritage. Nobody implemented that impulse more uncomplicatedly than the Mormons, a chosen people who believed the Lord when He told them to make the desert blossom as the rose. Nobody expressed it more bluntly than a Mormon hierarch, John Widtsoe, in the middle of the irrigation campaigns: "The destiny of man is to possess the whole earth; the destiny of the earth is to be subject to man. There can be no full conquest of the earth, and no real satisfaction to humanity, if large portions of the earth remain beyond his highest control" (*Success on Irrigation Projects,* p. 138).

That doctrine offends me to the bottom of my not-very-Christian soul. It is related to the spirit that builds castles of incongruous luxury in the desert. It is the same spirit that between 1930 and the present has so dammed, diverted, used, and reused the Colorado River that its saline waters now never reach the Gulf of California, but die in the sand miles from the sea; that has set the Columbia, a

45

far mightier river, to tamely turning turbines; that has reduced the Missouri, the greatest river on the continent, to a string of ponds; that has recklessly pumped down the water table of every western valley and threatens to dry up even so prolific a source as the Ogalalla Aquifer; that has made the Salt River Valley of Arizona, and the Imperial, Coachella, and great Central valleys of California into gardens of fabulous but deceptive richness; that has promoted a new rush to the West fated, like the beaver and grass and gold rushes, to recede after doing great environmental damage.

The Garden of the World has been a glittering dream, and many find its fulfillment exhilarating. I do not. I have already said that I think of the main-stem dams that made it possible as original sin, but there is neither a serpent nor a guilty first couple in the story. In Adam's fall we sinnéd all. Our very virtues as a pioneering people, the very genius of our industrial civilization, drove us to act as we did. God and Manifest Destiny spoke with one voice urging us to "conquer" or "win" the West; and there was no voice of comparable authority to remind us of Mary Austin's quiet but profound truth, that the manner of the country makes the usage of life there, and that the land will not be lived in except in its own fashion.

Obviously, reclamation is not the panacea it once seemed. Plenty of people in 1986 are opposed to more dams, and there is plenty of evidence against the long-range viability and the social and environmental desirability of large-scale irrigation agriculture. Nevertheless, millions of Americans continue to think of the water engineering in the West as one of our proudest achievements, a technology that we should export to backward Third World nations to help them become as we are. We go on praising apples as if eating them were an injunction of the Ten Commandments.

F O R its first thirty years, the Bureau of Reclamation struggled, plagued by money problems and unable to perform as its boosters had promised. It got a black eye for being involved, in shady ways, with William Mulholland's steal of the Owens Valley's water for the benefit of Los Angeles. The early dams it completed sometimes served not an acre of public land. It did increase homestead filings substantially, but not all those homesteads ended up in the hands of Jeffersonian yeomen: according to a 1922 survey, it had created few family farms; the 160-acre limitation was never enforced; three-quarters of the farmers in some reclamation districts were tenants.

Drought, the Great Depression, and the New Deal's effort to make public works jobs gave the bureau new life. It got quick appropriations for the building of the Boulder (Hoover) Dam, already authorized, and it took over from the state of California construction of the enormous complex of dams and ditches called the Central Valley Project, designed to harness all the rivers flowing westward out of the Sierra. It grew like a mushroom, like an exhalation. By the 1940s the bureau that only a few years before had been hanging on by a shoestring had built or was building the

four greatest dams ever built on earth up to that time—
Hoover, Shasta, Bonneville, and Grand Coulee—and
was already the greatest force in the West. It had discovered where power was, and allied itself with it: with the
growers and landowners, private and corporate, whose
interests it served, and with the political delegations, often
elected out of this same group, who carried the effort in
Washington for more and more pork barrel projects. In
matters of western water there are no political parties. You
cannot tell Barry Goldwater from Moe Udall, or Orrin
Hatch from Richard Lamm.

Nevertheless there was growing opposition to dams
from nature lovers, from economists and cost counters,
and from political representatives of areas that resented
paying these costs in subsidy of their competition. Unit-
ing behind the clause in the National Park Service Act
that enjoined "use without impairment," environmental
groups in 1955 blocked two dams in Dinosaur National
Monument and stopped the whole Upper Colorado River
Storage Project in its tracks. Later, in the 1960s, they also
blocked a dam in Marble Canyon, on the Colorado, and
another in Grand Canyon National Monument, at the
foot of the Grand Canyon.

In the process they accumulated substantial evidence,
economic, political, and environmental, against dams, the
bureau that built them, and the principles that guided that
bureau. President Jimmy Carter had a lot of public sympa-
thy when he tried to stop nine water-project boondoggles,
most of them in the West, in 1977. Though the hornet's

nest he stirred up taught him something about western water politics, observers noted that no new water projects were authorized by Congress until the very last days of the 99th Congress, in October 1986.

The great days of dam building are clearly over, for the best dam sites are used up, most of the rivers are "tamed," costs have risen exponentially, and public support of reclamation has given way to widespread and searching criticism. It is not a bad time to assess what the big era of water engineering has done to the West.

The voices of reappraisal are already a chorus. Four books in particular, all published within the past four years, have examined western water developments and practices in detail. They are Philip Fradkin's *A River No More,* about the killing of the Colorado; William A. Kahrl's *Water and Power,* on the rape of the Owens Valley by Los Angeles; Donald Worster's *Rivers of Empire,* a dismaying survey of our irrigation society in the light of Karl Wittvogel's studies of the ancient hydraulic civilizations of Mesopotamia and China; and Marc Reisner's *Cadillac Desert,* a history that pays particular unfriendly attention to the Bureau of Reclamation and its most empire-building director, Floyd Dominy.

None of those books is calculated to please agribusiness or the politicians and bureaucrats who have served it. Their consensus is that reclamation dams and their little brother the centrifugal pump have made an impressive omelet but have broken many eggs, some of them golden, and are in the process of killing the goose that laid them.

BEGIN with some environmental consequences of "taming" rivers, if only because the first substantial opposition to dams was environmental.

First, dams do literally kill rivers, which means they kill not only living water and natural scenery but a whole congeries of values associated with them. The scenery they kill is often of the grandest, for most main-stem dams are in splendid canyons, which they drown. San Francisco drowned the Hetch Hetchy Valley, which many thought as beautiful as Yosemite itself, to ensure its future water supply. Los Angeles turned the Owens Valley into a desert by draining off its natural water supplies. The Bureau of Reclamation drowned Glen Canyon, the most serene and lovely rock funhouse in the West, to provide peaking power for Los Angeles and the Las Vegas Strip.

The lakes formed behind dams are sometimes cited as great additions to public recreation, and Floyd Dominy even published a book to prove that the Glen Canyon Dam had beautified Glen Canyon by drowning it. But drawdown reservoirs rarely live up to their billing. Nothing grows in the zone between low-water mark and high-water mark, and except when brimming full, any drawdown reservoir, even Glen Canyon, which escapes the worst effects because its walls are vertical, is not unlike a dirty bathtub with a ring of mud and mineral stain around it.

A dammed river is not only stoppered like a bathtub, but it is turned on and off like a tap, creating a fluctuation of flow that destroys the riverine and riparian wildlife and creates problems for recreational boatmen who have to adjust to times when the river is mainly boulders and times when it rises thirty feet and washes their tied boats off the beaches. And since dams prohibit the really high flows of the spring runoff, boulders, gravel, and detritus pile up into the channel at the mouths of side gulches, and never get washed away.

Fishing too suffers, and not merely today's fishing but the future of fishing. Despite their fish ladders, the dams on the Columbia seriously reduced the spawning runs of salmon and steelhead, and they also trapped and killed so many smolts on their way downriver that eventually the federal government had to regulate the river's flow. The reduction of fishing is felt not only by the offshore fishing fleets and by Indian tribes with traditional or treaty fishing rights, but by sports fishermen all the way upstream to the Salmon River Mountains in Idaho.

If impaired rafting and fishing and sight-seeing seem a trivial price to pay for all the economic benefits supposedly brought by dams, reflect that rafting and fishing and sight-seeing are not trivial economic activities. Tourism is the biggest industry in every western state. The national parks, which are mainly in the public lands states, saw over three hundred million visitors in 1984. The national forests saw even more. A generation ago, only five thousand people in all the United States had ever rafted a river; by 1985, thirty-five million had. Every western river from the Rogue and the Owyhee to the Yampa, Green, San Juan, and Colorado is booked solid through the running season. As the rest of the country grows more stressful as a dwelling place, the quiet, remoteness, and solitude of a week on a wild river become more and more precious to more and more people. It is a good question whether we may not need that silence, space, and solitude for the healing of our raw spirits more than we need surplus cotton and alfalfa, produced for private profit at great public expense.

The objections to reclamation go beyond the obvious fact that reservoirs in desert country lose a substantial amount of their impounded water through surface evaporation; and the equally obvious fact that all such reservoirs eventually silt up and become mud flats ending in concrete waterfalls; and the further fact that an occasional dam, because of faulty siting or construction, will go out, as the Teton Dam went out in 1976, bringing disaster to people, towns, and fields below. They go beyond the fact that underground water, recklessly pumped, is quickly depleted, and that some of it will only be renewed in geological time, and that the management of underground water and that of surface water are necessarily linked. The ultimate objection is that irrigation agriculture itself, in deserts where surface evaporation is extreme, has a limited though unpredictable life. Marc Reisner predicts that in the next half century as much irrigated land will go out of production as the Bureau of Reclamation has "reclaimed" in its whole history.

Over time, salts brought to the surface by constant flooding and evaporation poison the soil: the ultimate, natural end of an irrigated field in arid country is an alkali flat. That was the end of fields in every historic irrigation civilization except Egypt, where, until the Aswan Dam, the annual Nile flood leached away salts and renewed the soil with fresh silt.

Leaching can sometimes be managed if you have enough sweet water and a place to put the runoff. But there is rarely water enough—the water is already 125 percent allocated and 100 percent used—and what water is available is often itself saline from having run through other fields upstream and having brought their salts back to the river. Colorado River water near the headwaters at Grand Lake is 200 parts per million (ppm) salt. Below the Wellton-Mohawk District on the Gila it is 6,300 ppm salt. The 1.5 million acre feet that we are pledged to deliver to Mexico is so saline that we are having to build a desalinization plant to sweeten it before we send it across the border.

Furthermore, even if you have enough water for occasional leaching, you have to have somewhere to drain off the waste water, which is likely not only to be saline, but to be contaminated with fertilizers, pesticides, and poisonous trace minerals such as selenium. Kesterson Reservoir, in the Central Valley near Los Banos, is a recent notorious instance, whose two-headed, three-legged, or merely dead waterfowl publicized the dangers of draining waste water off into a slough. If it is drained off into a river, or out to sea, the results are not usually so dramatic. But the inedible fish of the New River draining into the Salton Sea, and the periodically polluted beaches of Monterey Bay near the mouth of the Salinas River, demonstrate that agricultural runoff is poison anywhere.

The West's irrigated bounty is not forever, not on the scale or at the rate we have been gathering it in. The part of it that is dependent on wells is even more precarious than that dependent on dams. In California's San Joaquin Valley, streams and dams supply only 60 percent of the demand for water; the rest is pumped from wells—hundreds and thousands of wells. Pumping exceeds replenishment by a half-trillion gallons a year. In places the water table has been pumped down three hundred feet; in places the ground itself has sunk thirty feet or more. But with those facts known, and an end clearly in sight, nobody is willing to stop, and there is as yet no state regulation of groundwater pumping.

In Arizona the situation is if anything worse. 90 percent of Arizona's irrigation depends on pumping. And in Nebraska and Kansas and Oklahoma, old Dust Bowl country, they prepare for the next dust bowl, which is as inevitable as sunrise though a little harder to time, by pumping away the groundwater through center-pivot sprinklers.

Add to the facts about irrigation the fact of the oversubscribing of rivers. The optimists say that when more water is needed, the engineers will find a way—"augmentation" from the Columbia or elsewhere for the Colorado's overdrawn reservoirs, or the implementation of cosmic schemes such as NAWAPA (North American Water and Power Alliance), which would dam all the Canadian rivers up against the east face of the Rockies, and from that Mediterranean-sized reservoir supply water to every needy district from Minneapolis to Yuma. I think that there are geological as well as political difficulties in the way of water redistribution on that scale. The solution of western problems does not lie in more grandiose engineering.

Throw into the fact barrel, finally, a 1983 report from the Council on Environmental Quality concluding that desertification—the process of converting a viable arid-lands ecology into a lifeless waste—proceeds faster in the western United States than in Africa. Some of that desertification is the result of overgrazing, but the salinization of fields does its bit. When the hydraulic society falls back from its outermost frontiers, it will have done its part in the creation of new deserts.

T HE hydraulic society. I borrow the term from Donald Worster, who borrowed it from Karl Wittvogel. Wittvogel's studies convinced him that every hydraulic society is by necessity an autocracy. Power, he thought, inevitably comes to reside in the elite that understands and exercises control over water. He quotes C. S. Lewis: "What we call man's power over nature turns out to be a power exercised by some men over other men with nature as its instrument" (*The Abolition of Man,* p. 35); and Andre Gorz: "The total domination of nature inevitably entails a domination of people by the techniques of domination" (*Ecology as Politics,* p. 20). Those quotations suggest a very different approach from the human domination advocated by such as John Widtsoe.

The hydraulic society involves the maximum domination of nature. And the American West, Worster insists, is the greatest hydraulic society the world ever saw, far surpassing in its techniques of domination the societies on the Indus, the Tigris-Euphrates, or the Yellow River. The West, which Walter Webb and Bernard DeVoto both feared might remain a colonial dependency of the East, has instead become an empire and gotten the East to pay most of the bills.

The case as Worster puts it is probably overstated. There are, one hopes, more democratic islands than he allows for, more areas outside the domination of the water managers and users. Few parts of the West are totally controlled by what Worster sees as a hydraulic elite. Nevertheless, no one is likely to call the agribusiness West, with most of its power concentrated in the Iron Triangle of growers, politicians, and bureaucratic experts and its work done by a permanent underclass of dispossessed, mainly alien migrants, the agrarian democracy that the Newlands Act was supposed to create.

John Wesley Powell understood that a degree of land monopoly could easily come about in the West through control of water. A thorough Populist, he advocated cooperative rather than federal waterworks, and he probably never conceived of anything on the imperial scale later realized by the Bureau of Reclamation. But if he were alive today he would have to agree at least partway with Worster: water experts ambitious to build and expand their bureau and perhaps honestly convinced of the worth of what they are doing have allied themselves with landowners and politicians, and by making land monopoly through water control immensely profitable for their backers, they have made it inevitable.

How profitable? Worster cites figures from one of the most recent of the mammoth projects, the Westlands, that brought water to the western side of the San Joaquin Valley. Including interest over forty years, the cost to the taxpayers was $3 billion. Water is delivered to the beneficiaries, mostly large landholders, at $7.50 an acre foot—far below actual cost, barely enough to pay operation and maintenance costs. According to a study conducted by economists Philip LaVeen and George Goldman, the subsidy amounted to $2,200 an acre, $352,000 per quarter section—and very few quarter-section family farmers were among the beneficiaries. Large landholders obliged by the 160-acre limitation to dispose of their excess lands disposed of them to family members and cronies, paper farmers, according to a pattern by now well established among water users.

So much for the Jeffersonian yeoman and the agrarian democracy. As for another problem that Powell foresaw, the difficulty that a family would have in handling even 160 acres of intensively farmed irrigated land, both the corporate and the family farmers solve it the same way: with migrant labor, much of it illegally recruited below the Rio Grande. It is anybody's guess what will happen now that Congress has passed the Immigration Reform and Control Act, but up to now the border has been a sieve, carefully kept open from this side. On a recent rafting trip through the Big Bend canyons of the Rio Grande, my son twice surprised sheepdog functionaries herding wetbacks to safety in America.

Those wetbacks are visible not merely in California and

Texas, but pretty much throughout the West. Visiting Rigby, Idaho, up in the farming country below the washed-out Teton Dam, I found a shantytown where the universal language was Spanish. Wherever there are jobs to do, especially laborious or dirty jobs—picking crops, killing turkeys—there have been wetbacks brought in to do them. Like drug running, the importation of illegals has resulted from a strong, continuing American demand, most of it from the factories in the fields of the hydraulic society. One has to wonder if penalties for such importations will inhibit growers any more than the 160-acre limitation historically did.

M ARC Reisner, in *Cadillac Desert,* is less concerned with the social consequences than with the costs and environmental losses and the plain absurdities of our long battle with aridity.

> Only a government that disposes of a billion dollars every few hours would still be selling water in deserts for less than a penny a ton. And only an agency as antediluvian as the Bureau of Reclamation, hiding in a government as elephantine as ours, could successfully camouflage the enormous losses the taxpayer has to bear for its generosity. (P. 500)

Charles P. Berkey of Columbia University, a hydrologist, wrote in 1946,

The United States has virtually set up an empire on impounded and re-distributed water. The nation is encouraging development, on a scale never before attempted, of lands that are almost worthless except for the water that can be delivered to them by the works of man. There is building up, through settlement and new population, a line of industries foreign to the normal resources of the region. . . . One can claim (and it is true) that much has been added to the world; but the longer range view in this field, as in many others, is threatened by apparently incurable ailments and this one of slowly choking to death with silt is the most stubborn of all. There are no permanent cures. (Letter quoted in Kazmann, *Modern Hydrology,* p. 124)

Raphael Kazmann himself agrees:

The reservoir construction program, objectively considered, is really a program for the continued and endless expenditure of ever-increasing sums of public money to combat the effects of geologic forces, as these forces strive to reach positions of relative equilibrium in the region of rivers and the flow of water. It may be that future research in the field of modern hydrology will be primarily to find a method of extricating ourselves from this unequal struggle with minimum loss to the nation. (P. 125)

And Donald Worster pronounces the benediction: "The next stage after empire is decline." The West, aware of its own history, might phrase it differently: The next stage after boom is bust. Again.

WHAT should one make of facts as depressing as these? What do such facts do to the self-gratifying image of the West as the home of freedom, independence, largeness, spaciousness, and of the Westerner as total self-reliance on a white stallion? I confess they make this Westerner yearn for the old days on the Milk and the Missouri when those rivers ran free, and we were trying to learn how to live with the country, and the country seemed both hard and simple, and the world and I were young, when irrigation had not yet grown beyond its legitimate bounds and the West provided for its thin population a hard living but a wonderful life.

Sad to say, they make me admit, when I face them, that the West is no more the Eden that I once thought it than the Garden of the World that the boosters and engineers tried to make it; and that neither nostalgia nor boosterism can any longer make a case for it as the geography of hope.

. . . .

THREE

Variations on a Theme by Crèvecoeur

THERE are many kinds of wildernesses, Aldo Leopold wrote in *A Sand County Almanac,* and each kind forces on people a different set of adaptations and creates a different pattern of life, custom, and belief. These patterns we call cultures.

By that criterion, the West should have a different cultural look from other American regions, and within the regional culture there should be discernible a half dozen subcultures stemming from our adaptations to shortgrass plains, alpine mountains, slickrock canyons, volcanic scablands, and both high and low deserts.

But cultural differentiation takes a long time, and happens most completely in isolation and to homogeneous peoples, as it happened to the Paiutes. The West has had neither time nor isolation nor homogeneity of race and occupation. Change, both homegrown and imported, has overtaken time, time and again. We have to adapt not only to our changed physical environment but to our own adaptations, and sometimes we have to backtrack from our own mistakes.

Forming cultures involving heterogeneous populations do not grow steadily from definable quality to definable quality. Not only is their development complicated by class, caste, and social mobility, but they undergo simultaneous processes of erosion and deposition. They start from something, not from nothing. Habits and attitudes that have come to us embedded in our inherited culture, especially our inherited language, come incorporated in everything from nursery rhymes to laws and prayers, and they often have the durability of flint pebbles in puddingstone. No matter how completely their old matrix is dissolved, they remain intact, and are deposited almost unchanged in the strata of the new culture.

The population that for the eleven public lands states and territories was four million in 1900 was forty-five million in 1984, with at least a couple of million more, and perhaps twice that many, who weren't counted and didn't want to be. Many of those forty-five or forty-seven or forty-nine million came yesterday, since the end of World War II. They have not adapted, in the cultural sense, very far. Some of them are living anonymously in the Spanish-speaking barrios of San Diego, El Paso, Los Angeles, San Jose, where the Immigration Service can't find them. Some are experimenting with quick-change life-styles in the cultural confusion of western cities. Some are reading *Sunset Magazine* to find out what they should try to become. Some already know, from the movies and TV.

Being a Westerner is not simple. If you live, say, in Los Angeles, you live in the second largest city in the nation, urban as far as the eye can see in every direction except west. There is, or was in 1980—the chances would be somewhat greater now—a 6.6 percent chance that you are Asian, a 16.7 percent chance that you are black, and a 27 percent chance that you are Hispanic. You have only a 48 percent chance of being a non-Hispanic white.

This means that instead of being suitable for casting in the cowboy and pioneer roles familiar from the mythic and movie West, you may be one of those Chinks or Spics or Greasers for whom the legendary West had a violent contempt. You'd like to be a hero, and you may adopt the costume and attitudes you admire, but your color or language or the slant of your eyes tells you that you are one of the kind scheduled to be a villain or a victim, and your current status as second-class citizen confirms that view. You're part of a subculture envious of or hostile to the dominant one.

This ethnic and cultural confusion exists not only in Los Angeles but in varying proportions in every western city and many western towns. Much of the adaptation that is going on is adaptation to a very uncertain reality or to a reality whose past and present do not match. The western culture and western character with which it is easiest to identify exist largely in the West of make-believe, where they can be kept simple.

As invaders, we were rarely, or only temporarily, dependent on the materials, foods, or ideas of the regions we pioneered. The champagne and oysters that cheered midnight suppers during San Francisco's Gold Rush period were not local, nor was the taste that demanded them. The dominant white culture was always aware of its origins; it brought its origins with it across the plains or around the Horn, and it kept in touch with them.

The Spanish of New Mexico, who also brought their origins with them, are in other ways an exception. Settled at the end of the sixteenth century, before Jamestown and Quebec and well before the Massachusetts Bay Colony, New Mexico existed in isolation, dependent largely on

California, Central Valley Project, 1930s

itself, until the Americans forcibly took it over in 1846; and during those two and a half centuries it had a high Indian culture close at hand to teach it how to live with the country. Culturally, the Spanish Southwest is an island, adapted in its own ways, in many ways alien.

By contrast, the Anglo-American West, barely breached until the middle of the nineteenth century, was opened during a time of rapid communication. It was linked with the world by ship, rail, and telegraph before the end of the 1860s, and the isolation of even its brief, explosive outposts, its Alder Gulches and Cripple Creeks, was anything but total. Excited travelers reported the West in words to match its mountains; it was viewed in Currier and Ives prints drawn by enthusiasts who had never been there except in imagination. The outside never got over its heightened and romantic notion of the West. The West never got over its heightened and romantic notion of itself.

The pronounced differences that some people see between the West and other parts of America need to be examined. Except as they involve Spanish or Indian cultures, they could be mainly illusory, the result of the tendency to see the West in its mythic enlargement rather than as it is, and of the corollary tendency to take our cues from myths in the effort to enhance our lives. Life does sometimes copy art. More than drugstore cowboys and street corner Kit Carsons succumb. Plenty of authentic ranch hands have read pulp Westerns in the shade of the bunkhouse and got up walking, talking, and thinking like Buck Duane or Hopalong Cassidy.

No matter what kind of wilderness it developed in, every part of the real West was a melting-pot mixture of people from everywhere, operating under the standard American drives of restlessness, aggressiveness, and great expectations, and with the standard American freedom that often crossed the line into violence. It was supposed to be a democracy, and at least in the sense that it was often every man for himself, it was. Though some of its phases—the fur trade, the gold rushes, the open range cattle industry—lasted hardly longer than the blink of an eye, other phases—logging, irrigation farming, the stock farm with cattle or sheep—have lasted by now for a century or more, and have formed the basis for relatively stable communities with some of the attributes of place, some identity as subcultures of the prevailing postfrontier culture of America. If Turner's thesis is applicable beyond the 98th meridian, then the West ought to be, with minor local variations, America only more so.

Actually it is and it isn't. It would take fast footwork to dance the society based on big reclamation projects into a democracy. Even the cattle kingdom from which we derive our most individualistic and independent folk hero was never a democracy as the Middle West, say, was a democracy. The real-life cattle baron was and is about as democratic as a feudal baron. The cowboy in practice was and is an overworked, underpaid hireling, almost as homeless and dispossessed as a modern crop worker, and his fabled independence was and is chiefly the privilege of quitting his job in order to go looking for another just as bad. That, or go outside the law, as some did. There is a discrepancy between the real conditions of the West, which even among outlaws enforced cooperation and group effort, and the folklore of the West, which celebrated the dissidence of dissent, the most outrageous independence. Bernard DeVoto once cynically guessed that the only true individualists in the West wound up on the end of a rope whose other end was in the hands of a bunch of cooperators.

The dynamics of contemporary adaptation work ambiguously. The best imitators of frontier individualism these days are probably Silicon Valley and conglomerate executives, whose entrepreneurial attributes are not greatly different from those of an old-time cattle baron. Little people must salve with daydreams and fantasy the wounds of living. Some may imagine themselves becoming captains of industry, garage inventors whose inventions grow into Fortune 500 companies overnight; but I think that more of them are likely to cuddle up to a culture hero independent of the system and even opposed to it—a culture hero given them by Owen Wister, an eastern snob who saw in the common cowherd the lineaments of Lancelot. Chivalry, or the daydream of it, is at least as common among daydreamers as among entrepreneurs.

PHYSICALLY, the West could only be itself. Its scale, its colors, its landforms, its plants and animals, tell a traveler what country he is in, and a native that he is at home. Even western cities owe most of their distinctiveness to their physical setting. Albuquerque with its mud-colored houses spreading like clay banks along the valley of the Rio Grande could only be New Mexico. Denver's ringworm suburbs on the apron of the Front Range could only be boom-time Colorado. Salt Lake City bracing back against the Wasatch and looking out toward the dead sea and the barren ranges could only be the Great Basin.

But is anything except their setting distinctive? The people in them live on streets named Main and State, Elm and Poplar, First and Second, like Americans elsewhere. They eat the same Wheaties and Wonder Bread and Big Macs, watch the same ball games and soaps and sitcoms on TV, work at the same industrial or service jobs, suffer from the same domestic crises and industrial blights, join the same health clubs and neighborhood protective associations, and in general behave and misbehave much as they would in Omaha or Chicago or East Orange. The homogenizing media have certainly been at work on them, perhaps with more effect than the arid spaciousness of the region itself, and while making them more like everybody else have also given them misleading clues about who they are.

"WHO is the American, this new man?" Crèvecoeur asked rhetorically in his *Letters from an American Farmer* more than two hundred years ago, and went on to idealize him as the American farmer—industrious, optimistic, upwardly mobile, family-oriented, socially responsible, a new man given new hope in the new world, a lover of both hearth and earth, a builder of communities. He defined him in the terms of a new freedom, emancipated from feudalism, oppression, and poverty, but with no wish to escape society or its responsibilities. Quite the contrary.

Crèvecoeur also sketched, with distaste, another kind of American, a kind he thought would fade away with the raw frontier that had created him. This kind lived alone or with a slattern woman and a litter of kids out in the woods. He had no fixed abode, tilled no ground or tilled it only fitfully, lived by killing, was footloose, uncouth, anti-social, impatient of responsibility and law. The eating of wild meat, Crèvecoeur said, made him ferocious and gloomy. Too much freedom promoted in him a coarse selfishness and a readiness to violence.

The pioneer farmer as Crèvecoeur conceived him has a place in western history, and as the Jeffersonian yeoman he had a prominent place in the mistaken effort to oversettle the West, first by homestead and later by reclamation. Traces of him are to be found in western literature, art, and myth. Sculptors have liked his sturdy figure plodding beside the covered wagon on which ride his poke-bonneted wife and his barefoot children. He strides through a lot of WPA murals. The Mormons, farmers in the beginning, idealize him. He has achieved more than life size in such novels of the migration as *The Covered Wagon* and *The Way West*.

But those, as I have already suggested, are novels more of motion than of place, and the emigrants in them are simply farmer-pioneers on their way to new farms. They have not adapted to the West in the slightest degree. They belong where the soil is deep, where the Homestead Act worked, where settlers planted potato peelings in their fireguards and adjourned to build a combination school–church–social hall almost before they had roofs on their shanties. The pioneer farmer is a midwestern, not a western, figure. He is a pedestrian, and in the West, horseman's country even for people who never got on a horse in their lives, pedestrians suffer from the horseman's contempt that seems as old as the Scythians. The farmer's very virtues as responsible husband, father, and home builder are against him as a figure of the imagination. To the fantasizing mind he is dull, the ancestor of the clodhopper, the hayseed, and the hick. I have heard Wyoming ranch hands jeer their relatives from Idaho, not because the relatives were Mormons—so were the ranch hands—but because they were farmers, potato diggers.

It was Crèvecoeur's wild man, the borderer emancipated into total freedom, first in eastern forests and then in the plains and mountains of the West, who really fired our imaginations, and still does. We have sanitized him somewhat, but our principal folk hero, in all his shapes good and bad, is essentially antisocial.

In real life, as Boone, Bridger, Jed Smith, Kit Carson, he appeals to us as having lived a life of heroic courage, skill, and self-reliance. Some of his manifestations such as Wild Bill Hickok and Buffalo Bill Cody are tainted with outlawry or showmanship, but they remain more than life-size. Even psychopathic killers such as Billy the Kid and Tom Horn throw a long shadow, and some outlaws such as Butch Cassidy and Harry Longabaugh have all the engaging imitability of Robin Hood. What charms us in them is partly their daring, skill, and invulnerability, partly their chivalry; but not to be overlooked is their impatience with all restraint, their freedom from the social responsibility that Crèvecoeur admired in his citizen farmer, and that on occasion bows the shoulders of every man born.

Why should I stand up for civilization? Thoreau asked a lecture audience. Any burgher or churchwarden would stand up for that. Thoreau chose instead to stand up for wildness and the savage heart.

We all know that impulse. When youths run away from home, they don't run away to become farmers. They run away to become romantic isolates, lone riders who slit their eyes against steely distance and loosen the carbine in its scabbard when they see law, or obligation, or even company, approaching.

Lawlessness, like wildness, is attractive, and we conceive the last remaining home of both to be the West. In a folklore predominantly masculine and macho, even women take on the look. Calamity Jane is more familiar to us than Dame Shirley, though Dame Shirley had it all over Jane in brains, and could have matched her in courage, and lived in mining camps every bit as rough as the cow towns and camps that Calamity Jane frequented.

The attraction of lawlessness did not die with the frontier, either. Look at the survivalist Claude Dallas, who a couple of years ago killed two Idaho game wardens when they caught him poaching—shot them and then finished them off with a bullet in the back of the head. In that act of unchivalrous violence Dallas was expressing more than an unwillingness to pay a little fine. For months, until he was captured early in 1987, he hid out in the deserts of Idaho and Nevada, protected by people all over the area. Why did they protect him? Because his belated frontiersman style, his total self-reliance and physical competence, his repudiation of any control, appealed to them more than murder repelled them or law enlisted their support.

All this may seem remote from the life of the average Westerner, who lives in a city and is more immediately concerned with taxes, schools, his job, drugs, the World Series, or even disarmament, than with archetypal figures out of folklore. But it is not so remote as it seems. Habits persist. The hoodlums who come to San Francisco to beat up gays are vigilantes, enforcing their prejudices with violence, just as surely as were the miners who used to hunt down Indians and hang Chinese in the Mother Lode, or the ranchers who rode out to exterminate the nesters in Wyoming's Johnson County War.

Habits persist. The hard, aggressive, single-minded energy that according to politicians made America great is demonstrated every day in resource raids and leveraged takeovers by entrepreneurs; and along with that competitive individualism and ruthlessness goes a rejection of any controlling past or tradition. What matters is here, now, the seizable opportunity. "We don't need any history," said one Silicon Valley executive when the Santa Clara County Historical Society tried to bring the electronics industry together with the few remaining farmers to discuss what was happening to the valley that only a decade or two ago was the fruit bowl of the world. "What we need is more attention to our computers and the moves of the competition."

We are not so far from our models, real and fictional, as we think. As on a wild river, the water passes, the waves remain. A high degree of mobility, a degree of ruthlessness, a large component of both self-sufficiency and self-righteousness mark the historical pioneer, the lone-riding folk hero, and the modern businessman intent on opening new industrial frontiers and getting his own in the process. The same qualities inform the extreme individualists who believe that they belong to nothing except what they choose to belong to, those who try on life-styles as some try on clothes, whose only communal association is with what Robert Bellah calls "life-style enclaves," casual and temporary groupings of the like-minded. One reason why it is so difficult to isolate any definitely western culture is that so many Westerners, like other Americans only more so, shy away from commitment. Mobility of every sort—physical, familial, social, corporate, occupational, religious, sexual—confirms and reinforces the illusion of independence.

Back to the freedom-loving loner, whom we might call Leatherstocking's descendant, as Henry Nash Smith taught us to, if all that tribe were not childless as well as orphaned. In the West this figure acquired an irresistible costume—the boots, spurs, chaps, and sombrero bequeathed to him by Mexican vaqueros, plus the copper-riveted canvas pants invented for California miners by a peddler named Levi Strauss—but he remained estranged from real time, real place, and any real society or occupation. In fact, it is often organized society, in the shape of a crooked sheriff and his cronies, that this loner confronts and confounds.

The notion of civilization's corruption, the notion that the conscience of an antisocial savage is less calloused than the conscience of society, is of course a bequest from Jean-Jacques Rousseau. The chivalry of the antisocial one, his protectiveness of the weak and oppressed, especially those whom James Fenimore Cooper customarily referred to as "females," is from Cooper with reinforcement from two later romantics, Frederic Remington and Owen Wister, collaborators in the creation of the knight-errant in chaps.

The hero of Wister's 1902 novel *The Virginian* is gentle-seeming, easygoing, humorous, but when the wicked force him into action he is the very gun of God, better at violence than the wicked are. He is a daydream of glory made flesh. But note that the Virginian not only defeats Trampas in a gunfight as formalized as a fourteenth-century joust, the first of a thousand literary and movie walk-downs, but he also joins the vigilantes and in the name of law and order acts as jury, judge, and hangman for his friend Shorty, who has gone bad and become a rustler.

The Virginian feels sorry about Shorty, but he never questions that the stealing of a few mavericks should be punished by death, any more than Wister questioned the motives of his Wyoming rancher host who led the Johnson County vigilantes against the homesteaders they despised and called rustlers. This culture hero is himself law. Law is whatever he and his companions (and employers) believe (which means law is his and their self-interest).

Whatever action he takes is law enforcement. Compare Larry McMurtry's two ex-Texas Rangers in *Lonesome Dove*. They kill more people than all the outlaws in that book put together do, but their killings are *right*. Their lawlessness is justified by the lack of any competing socialized law, and by a supreme confidence in themselves, as if every judgment of theirs could be checked back to Coke and Blackstone, if not to Leviticus.

Critics have noted that in *The Virginian* (and for that matter in most of its successors, though not in *Lonesome Dove*) there are no scenes involving cattle. There is no manure, no punching of postholes or stringing of barbed wire, none of the branding, castrating, dehorning, dipping, and horseshoeing that real cowboys, hired men on horseback, spend their laborious and unromantic lives at. The physical universe is simplified like the moral one. Time is stopped.

The Virginian is the standard American orphan, dislocated from family, church, and place of origin, with an uncertain past identified only by his nickname. With his knightly sense of honor and his capacity to outviolence the violent, he remains an irresistible model for romantic adolescents of any age, and he transfers readily from the cowboy setting to more modern ones. It is only a step from his "When you call me that, smile," to the remark made famous by the current mayor of Carmel and the fortieth president of the United States: "Go ahead, make my day."

There are thousands more federal employees in the West than there are cowboys—more bookkeepers, aircraft and electronics workers, auto mechanics, printers, fry cooks. There may be more writers. Nevertheless, when most Americans east of the Missouri hear the word "West" they think "cowboy." Recently a documentary filmmaker asked me to be a consultant on a film that would finally reveal the true West, without romanticizing or adornment. It was to be done by chronicling the life of a single real-life individual. Guess who he was. A cowboy, and a rodeo cowboy at that—a man who had run away from his home in Indiana at the age of seventeen, worked for a year on a Texas ranch, found the work hard, made his way onto the rodeo circuit, and finally retired with a lot of his vertebrae out of line to an Oklahoma town where he made silver-mounted saddles and bridles suitable for the Sheriff's Posse in a Frontier Days parade, and spun yarns for the wide-eyed local young.

Apart from the fantasy involved in it, which is absolutely authentic, that show business life is about as typically western as a bullfighter's is typically Spanish. The critics will probably praise the film for its realism.

I spend this much time on a mythic figure who has irritated me all my life because I would obviously like to bury him. But I know I can't. He is a faster gun than I am. He is too attractive to the daydreaming imagination. It gets me nowhere to object to the self-righteous, limited, violent code that governs him, or to disparage the novels of Louis L'Amour because they are mass-produced with interchangeable parts. Mr. L'Amour sells in the millions, and has readers in the White House.

But what one can say, and be sure of, is that even while

the cowboy myth romanticizes and falsifies western life, it says something true about western, and hence about American, character.

Western culture and character, hard to define in the first place because they are only half-formed and constantly changing, are further clouded by the mythic stereotype. Why hasn't the stereotype faded away as real cowboys became less and less typical of western life? Because we can't or won't do without it, obviously. But also there is the visible, pervasive fact of western space, which acts as a preservative. Space, itself the product of incorrigible aridity and hence more or less permanent, continues to suggest unrestricted freedom, unlimited opportunity for testings and heroisms, a continuing need for self-reliance and physical competence. The untrammeled individualist persists partly as a residue of the real and romantic frontiers, but also partly because runaways from more restricted regions keep reimporting him. The stereotype continues to affect romantic Westerners and non-Westerners in romantic ways, but if I am right it also affects real Westerners in real ways.

In the West it is impossible to be unconscious of or indifferent to space. At every city's edge it confronts us as federal lands kept open by aridity and the custodial bureaus; out in the boondocks it engulfs us. And it does contribute to individualism, if only because in that much emptiness people have the dignity of rareness and must do much of what they do without help, and because self-reliance becomes a social imperative, part of a code. Witness the crudely violent code that governed a young Westerner like Norman Maclean, as he reported it in the stories of *A River Runs through It*. Witness the way in which space haunts the poetry of such western poets as William Stafford, Richard Hugo, Gary Snyder. Witness the lonely, half-attached childhood of a writer such as Ivan Doig. I feel the childhood reported in his *This House of Sky* because it is so much like my own.

Even in the cities, even among the dispossessed migrants of the factories in the fields, space exerts a diluted influence as illusion and reprieve. Westerners live outdoors more than people elsewhere, because outdoors is mainly what they've got. For clerks and students, factory workers and mechanics, the outdoors is freedom, just as surely as it is for the folkloric and mythic figures. They don't have to own the outdoors, or get permission, or cut fences, in order to use it. It is public land, partly theirs, and that space is a continuing influence on their minds and senses. It encourages a fatal carelessness and destructiveness because it seems so limitless and because what is everybody's is nobody's responsibility. It also encourages, in some, an impassioned protectiveness: the battlegrounds of the environmental movement lie in the western public lands. Finally, it promotes certain needs, tastes, attitudes, skills. It is those tastes, attitudes, and skills, as well as the prevailing destructiveness and its corrective, love of the land, that relate real Westerners to the myth.

David Rains Wallace, in *The Wilder Shore*, has traced the effect of the California landscape—the several California landscapes from the Pacific shore to the inner deserts—on

California writers. From Dana to Didion, the influence has been varied and powerful. It is there in John Muir ecstatically riding a storm in the top of a two-hundred-foot sugar pine; in Mary Austin quietly absorbing wisdom from a Paiute basketmaker; in Jack London's Nietzschean supermen pitting themselves not only against society but against the universe; in Frank Norris's atavistic McTeague, shackled to a corpse that he drags through the 130-degree heat of Death Valley; and in Robinson Jeffers on his stone platform between the stars and the sea, falling in love outward toward space. It is also there in the work of western photographers, notably Ansel Adams, whose grand, manless images are full of the awe men feel in the face of majestic nature. Awe is common in that California tradition. Humility is not.

Similar studies could be made, and undoubtedly will be, of the literature of other parts of the West, and of special groups of writers such as Native Americans who are mainly western. The country lives, still holy, in Scott Momaday's *Way to Rainy Mountain*. It is there like a half-forgotten promise in Leslie Marmon Silko's *Ceremony,* and like a homeland lost to invaders in James Welch's *Winter in the Blood* and Louise Erdrich's *Love Medicine*. It is a dominating presence, as I have already said, in the work of Northwest writers.

Western writing turns out, not surprisingly, to be largely about things that happen outdoors. It often involves characters who show a family resemblance of energetic individualism, great physical competence, stoicism,

determination, recklessness, endurance, toughness, re-
belliousness, resistance to control. It has, that is, residual
qualities of the heroic, as the country in which it takes
place has residual qualities of the wilderness frontier.

Those characteristics are not the self-conscious creation
of regional patriotism, or the result of imitation of older
by younger, or greater by lesser, writers. They are ines-
capable; western life and space generate them; they are
what the faithful mirror shows. When I wrote *The Big
Rock Candy Mountain* I was ignorant of almost everything
except what I myself had lived, and I had no context for
that. By the time I wrote *Wolf Willow,* a dozen years later,
and dealt with some of the same experience from another
stance, I began to realize that my Bo Mason was a character with relatives throughout western fiction. I could see
in him resemblance to Ole Rölvaag's Per Hansa, to Mari
Sandoz's Old Jules, to A. B. Guthrie's Boone Caudill,
even to the hard-jawed and invulnerable heroes of the
myth. But I had not been copying other writers. I had
been trying to paint a portrait of my father, and it happened that my father, an observed and particular individual, was also a type—a very western type.

Nothing suggests the separateness of western experience so clearly as the response to it of critics nourished in
the Europe-oriented, politicized, sophisticated, and anti-
heroic tradition of between-the-wars and postwar New
York. Edmund Wilson, commenting on Hollywood writers, thought of them as wreathed in sunshine and
bougainvillea, "spelling cat for the unlettered"; or as senti-

mental toughs, the boys in the back room; or as easterners of talent (Scott Fitzgerald was his prime example) lost to significant achievement and drowning in the La Brea tar pits.

Leslie Fiedler, an exponent of the *Partisan Review* subculture, came west to teach in Missoula in the 1950s and discovered "the Montana face"—strong, grave, silent, bland, untroubled by thought, the face of a man playing a role invented for him two centuries earlier and a continent-and-ocean away by a French romantic philosopher.

Bernard Malamud, making a similar pilgrimage to teach at Oregon State University in Corvallis, found the life of that little college town intolerable, and retreated from it to write it up in the novel *A New Life*. His Gogolian antihero S. Levin, a Jewish intellectual, heir to a thousand years of caution, deviousness, spiritual subtlety, and airless city living, was never at home in Corvallis. The faculty he was thrown among were suspiciously open, overfriendly, overhearty, outdoorish. Instead of a commerce in abstract ideas, Levin found among his colleagues a devotion to the art of fly-fishing that simply bewildered him. Grown men!

If he had waited to write his novel until Norman Maclean had written the stories of *A River Runs through It*, Malamud would have discovered that fly-fishing is not simply an art but a religion, a code of conduct and language, a way of telling the real from the phony. And if Ivan Doig had written before Leslie Fiedler shook up Missoula by the ears, Fiedler would have had another view of the Montana face. It looks different, depending on whether you encounter it as a bizarre cultural artifact on a Montana railroad platform, or whether you see it as young Ivan Doig saw the face of his dependable, skilled, likable, rootless sheepherder father. Whether, that is, you see it from outside the culture, or from inside.

In spite of the testimony of Fiedler and Malamud, if I were advising a documentary filmmaker where he might get the most quintessential West in a fifty-six-minute can, I would steer him away from broken-down rodeo riders, away from the towns of the energy boom, away from the cities, and send him to just such a little city as Missoula or Corvallis, some settlement that has managed against difficulty to make itself into a place and is likely to remain one. It wouldn't hurt at all if this little city had a university in it to keep it in touch with its cultural origins and conscious of its changing cultural present. It would do no harm if an occasional Leslie Fiedler came through to stir up its provincialism and set it to some self-questioning. It wouldn't hurt if some native-born writer, some Doig or Hugo or Maclean or Welch or Kittredge or Raymond Carver, were around to serve as culture hero—the individual who transcends his culture without abandoning it, who leaves for a while in search of opportunity and enlargement but never forgets where he left his heart.

It is in places like these, and through individuals like these, that the West will realize itself, if it ever does: these towns and cities still close to the earth, intimate and interdependent in their shared community, shared optimism, and shared memory. These are the seedbeds of an emergent western culture. They are likely to be there when the agribusiness fields have turned to alkali flats and the dams have silted up, when the waves of overpopulation that have been destroying the West have receded, leaving the stickers to get on with the business of adaptation.

BIBLIOGRAPHY

. . . .

The titles listed here are only those that were most immediately useful in the preparation of these lectures. They do not constitute more than the minutest fraction of a comprehensive bibliography of the West, but all of them would belong on such a comprehensive list, and most of them would be indispensable.

Austin, Mary. *The Land of Little Rain.* Boston and New York, 1903.

Bellah, Robert; Madsen, Richard; Sullivan, William M.; Swidler, Ann; and Tipton, Steven M. *Habits of the Heart.* Berkeley, 1985.

Crèvecoeur, Hector St. John. *Letters from an American Farmer,* edited with a preface by William P. Trent and an introduction by Ludwig Lewisohn. New York, 1925.

DeVoto, Bernard. *The Uneasy Chair.* Boston, 1955.

Doig, Ivan. *This House of Sky.* New York, 1978.

Fradkin, Philip L. *A River No More: The Colorado River and the West.* New York, 1981.

Gilpin, William. *The Mission of the North American People—Geographical, Social, and Political.* Philadelphia, 1873.

Kahrl, William A. *Water and Power*. Berkeley, 1982.

Kazmann, Raphael. *Modern Hydrology*. 2d ed. New York, 1972.

Leopold, Aldo. *A Sand County Almanac and Sketches Here and There*. New York, 1949.

Maclean, Norman. *A River Runs through It*. Chicago and London, 1976.

Moon, William Least Heat. *Blue Highways*. New York, 1984.

Powell, J. W. *Report on the Lands of the Arid Region of the United States, with a More Detailed Account of the Lands of Utah*. 45th Cong., 2d sess., 1878. H.R. Exec. Doc. 73.

Reisner, Marc. *Cadillac Desert: The American West and Its Disappearing Water*. New York, 1986.

Smith, Henry Nash. *Virgin Land*. Cambridge, 1950.

Stegner, Wallace. *Beyond the Hundredth Meridian: John Wesley Powell and the Second Opening of the West*. Boston, 1954.

———. *Wolf Willow: A History, a Story, and a Memory of the Last Plains Frontier*. New York, 1962.

Turner, Frederick Jackson. *The Frontier in American History*. New York, 1920.

Wallace, David Rains. *The Wilder Shore*. San Francisco, 1985.

Webb, Walter Prescott. *The Great Plains*. New York, 1931.

White, Lynn. "Historical Roots of Our Ecologic Crisis." *Science* 155(March 10, 1967):1203–7.

Widtsoe, John. *Success on Irrigation Projects*. New York, 1928.

Wister, Owen. *The Virginian*. New York, 1902.

Worster, Donald. *Rivers of Empire: Water, Aridity, and the Growth of the American West*. New York, 1985.

Zaslowsky, Dyan, and the Wilderness Society. *These American Lands*. New York, 1986.